S. Edward (Samuel Edward) Warren

Notes on Polytechnic or Scientific Schools in the United

States

Their Nature, Position, Aims and Wants

.

S. Edward (Samuel Edward) Warren

Notes on Polytechnic or Scientific Schools in the United States
Their Nature, Position, Aims and Wants

ISBN/EAN: 9783743687028

Printed in Europe, USA, Canada, Australia, Japan

Cover: Foto ©berggeist007 / pixelio.de

More available books at **www.hansebooks.com**

NOTES

ON

𝔓𝔬𝔩𝔶𝔱𝔢𝔠𝔥𝔫𝔦𝔠 𝔬𝔯 𝔖𝔠𝔦𝔢𝔫𝔱𝔦𝔣𝔦𝔠 𝔖𝔠𝔥𝔬𝔬𝔩𝔰,

IN THE

UNITED STATES;

THEIR NATURE, POSITION, AIMS AND WANTS.

"Is there any such happiness as for a man's mind to be raised above the confusion of things, where he may have the prospect of the order of nature?"

"Are we the richer, by one poor invention, by reason of all the learning that hath been these many hundred years?"

BY S. EDWARD WARREN, C. E.,

PROF. OF DESCRIPTIVE GEOMETRY, ETC. IN THE RENSSELAER POLYTECHNIC INSTITUTE, AND GRADUATE OF THE SAME. (CLASS OF '51.)

NEW YORK:
JOHN WILEY AND SON, 535 BROADWAY.
1866.

PREFATORY NOTE.

These unpretending pages, put forth in advance of a possible fuller treatment of their subject, are an attempt to respond, even if but very briefly, and provisionally, to much earnest inquiry concerning the true nature, position, and aims of Polytechnic Schools; and to the evident immediate need of correct popular information relative to them. It is hoped that they may also contribute to unity of sentiment and action, both among their friends in the community at large, especially their alumni, and among their officers and thoughtful and earnest members. That the need just alluded to exists, is not surprising. The whole class of Polytechnic—otherwise called Scientific, Technical, Technological, or Industrial—Schools is of modern origin everywhere, and in this country, comparatively unique. Hence misapprehension of their true nature and grade, and consequent legitimate mode of administration, not unnaturally arises, on slight misleading occasions.

For statements of facts, we have relied on official publications, correspondence, and standard educational literature, without, however, interrupting the reader by continual foot note references to them. The statistics of the concluding section are mainly abridged from the valuable report on the reorganization and proposed development of the Rensselaer Polytechnic Institute, prepared in 1855, by the then Director.* We have been unable, in the short time which could be spared for recording these notes, to hunt up many later or fuller authorities.

JANUARY, 1866.

---

* B. FRANKLIN GREENE, A. M., C. E.

NOTES ON POLYTECHNIC SCHOOLS.

I.

A List of such Schools in the United States, With Brief Explanatory Notes.

It will be convenient to present, first, in these notes, a list of the existing "Scientific Departments" and Technical Schools in the United States, so far as they are known to the writer.

In reference to the first section of the following table, it should be understood that it embraces those schools, whose character as truly distinct professional schools, is most apparent. This distinctive character is more or less obscured in the case of the schools named in the second part of the table, owing to their *comparatively undeveloped condition*, so far as now known, or else to their being *merged in the general courses of the institutions including them.* Hence it has been impossible to arrange them in the same list with those of the first section, in the order of definite dates of beginning.

The familiar professional schools—Theological, Medical and Legal—are, as is well known, sometimes separate institutions, and sometimes, attached to colleges. The same is true of Scientific Professional Schools. Hence, in any case where the name does not indicate the fact, the first column of the table shows the condition, in this respect, of each school mentioned.

TABLE OF POLYTECHNIC SCHOOLS, SCIENTIFIC DEPARTMENTS, ETC., IN THE UNITED STATES.

SECTION I.

NAME AND CONDITION.	COMPONENT SCHOOLS, (Or Courses.)	LENGTH OF COURSES.	LOCATION.	DATE.	TOTAL ATTEND'NCE.	AGE FOR ADMISSION.	DEGREES CONFERRED.
1—RENSSELAER POLYTECHNIC INSTITUTE. (Independent.)	Civil Engineering. Mechanical Engineering. Topographical Engineering. Natural Science.	4 years. " " "	Troy, N.Y.	1824.	150 in 1865	16.	C.E. M.E. T.E. B.S.
2—SCIENTIFIC COURSE AT UNION COLLEGE.	Applied Chemistry. Engineering.	2 years. "	Schenectady, N.Y.	1845.	40 in 1865	16.	C.E.
3—SHEFFIELD SCIENTIFIC SCHOOL. (One of the Professional Schools around Yale College.)	General Course. Chemistry and Natural Science. Engineering. Engineering—Higher. Agriculture and Mechanic Arts.	3 years. 3 " 2 " 3 "	New Haven, Conn.	1847.	57 in 1864	16.	Bach. of Phil. " " C.E.
4—LAWRENCE SCIENTIFIC SCHOOL. (One of the Professional Schools around Harvard College.)	Chemistry, General & Technical. Zoology and Geology. Engineering. Botany. Compar. Anatomy & Physiology. Mineralogy.	At least 1 year. " " " "	Cambridge, Mass.	1848.	75 in 1864	18.	B.S. " " " "
5—POLYTECHNIC COLLEGE OF PENNSYLVANIA. (Independent.)	General School. Civil Engineering. Mechanical Engineering. Practical Chemistry. Agriculture. Mines. Architecture.	1 year. 2 years. " " " "	Philadelphia, Pa.	1851.	136 in 1865	16.	Bach. of Civ. Eng. " of Mech. Eng. " of Chem. Eng. " of Agrl. Eng. " of Min. Eng. " of Arch.
6—CHANDLER SCIENTIFIC SCHOOL. (One of the Professional Schools around Dartmouth College.)	General Course. Engineering Course. Commercial Course. Higher General Course.	3 years. 1 year. 1 " 1	Hanover, N.H.	1852.	43 in 1865	Not stated.	B.S.
7—SCIENTIFIC COURSES IN UNIVERSITY OF MICHIGAN. (General Scientific Course 4 years.)	Civ. Eng. in Gen'l Sci. Course. " " in Special Course. Natural History. General & Technical Chemistry. Mines.	3 years. 1 year. Indefinite "	Ann Arbor, Mich.	1850 to 1856.	82 in 1865 139	Not stated.	C.E. B.S. M.E.

Institution	Courses / Departments	Duration	Location	Established			Degrees
8—"PROFESSIONAL (SCIENTIFIC) DEPARTMENT." (Among the Prof'l Schools of Univ. of N. Y. City.)	Civil Eng. & Architeture, Analyt. & Practical Chemistry.	3 years. 2 "	New York City.	1856.	31 in 1865	Not stated.	C. E. B. S. & Ph. D.
9—SCIENTIFIC SECTION OF WASHINGTON UNIVERSITY.	Gen'l Scien. & Tech. Course. O'FALLON POLYTECH. INSTITUTE. {a—Evening School. b—Popular Scien. Lectures c—School of Design. d—School of Mathematics.	3 years. 3 years.	St. Louis, Mo.	1857. Not in full operation.	7 in 1864	16.	B. S.
10—COOPER UNION FOR THE ADVANCEMENT OF SCIENCE AND ART. (Independent.)	Free Night School of Science. " " Art. School of Design for Women. Scientific (Polytechnic Day) School.	5 years. Not over 4 years.	New York City.	1859. 1859. Not yet established.	1231 in 1864 173 in 1864	16.	Diploma & Medal. Degrees.
11—COLLEGIATE AND ENGINEERING INSTITUTE. (Independent.)		2 years.	New York City.	1862.		Not stated.	Diploma.
12—SCHOOL OF MINES. (Of Columbia College.)		3 years.	New York City.	1864.	33 in 1865	16.	M. E., Ph. B., and Ph. D.
13—MASSACHUSETTS INSTITUTE OF TECHNOLOGY. (Full Course 4 years.—Independent.)	General Course, common to all the following Special Courses: {Mechanical Engineering. Civil & Topographical Eng. Practical Chemistry. Geology and Mining. Building and Architecture. Higher General Science, etc.	2 years. 2 years.	Boston, Mass.	1865.	72 in 1865	16.	Degrees in these several special Courses.
14—WORCESTER CO. FREE INDUSTRIAL INSTITUTE. (Independent.)	Not yet in operation.		Worcester, Mass.	1865.			
15—CORNELL UNIVERSITY.	Not yet in operation.		Ithaca, N. Y.	1865.			
16—UNIVERSITY OF THE SOUTH. (Not yet in operation.)	School of Civil Engineering, " " Technical Chemistry. " " Agriculture. " " Commerce and Trade. " " Mines.		Sewanee, Tenn.	About 1860.			
17—DELAWARE LITERARY INSTITUTE AND ENGINEERING SCHOOL.			Franklin, N. Y.	1865.			
18—SCHOOL OF MINES. (In connection with Harvard College.)			Cambridge, Mass.	1865.			

SECTION II:

NAMES.	LOCATION.	FOUNDED IN :	ATTENDANCE.
Brown University. Dep't. of Chemistry and Engineering,	Providence, R. I.		
New York Free Academy,	N. Y. City,	1853,	482 in 1864.
Brooklyn Collegiate and Polytechnic Institute,	Brooklyn, N. Y.,	1855,	
University of Pennsylvania, Dep't. of Mines, Arts and Manufactures.	Philadelphia, Pa.		
Washington College. Dep't. of Practical Mechanics.	Lexington, Va.	1866.	

Some of the institutions named in the foregoing table, are characterized by distinctive features, so marked and peculiar, that a brief mention of them is added, so far as it may favor a fuller understanding of the table. It is, however, as foreign to our purpose, as to our place, to offer critical notices, or enter into comparisons, at least in case of actually existing institutions.

THE RENSSELAER POLYTECHNIC INSTITUTE is distinguished as the pioneer of its class in this country. At first, more known as a school of Technical *Natural* Science, than of late years ; its present character, to which it owes so much of its prestige, was impressed upon it during a transition period of about five years, beginning in 1849. If it be added that, until within a very brief period, it stood alone in respect to the extent and elevation of its curriculum, it is saying no more than ought to be true of the senior institution. The assertion is also justified by the facts : first, that most of its graduates, of late years, have required the full four years for the completion of their course of study ; and second, that, nevertheless, the average age of its first year men, or " Division D," has been from one to two years above the minimum age (sixteen) for admission, while the average age of its present second year men, or "Division C," of over fifty members, is scarcely less than *three* years above its corresponding minimum required age, (seventeen).

THE SHEFFIELD SCIENTIFIC SCHOOL is in part characterized by its connection with Yale College, which has long been a distinguished home for the culture of the Natural Sciences.

THE LAWRENCE SCIENTIFIC SCHOOL possesses a distinctive peculiarity of organization, by which limited fields of study are marked out as departments, which are kept so far distinct, that separate arrangements, as to tuition fees and times of instruction, are required for each.

THE COOPER UNION is distinguished by its character as a most noble charity, bright, it is not too much to say, in the constellation of the world's best charities—charities of that nature that it is no humiliation at all, but a high honor, to be intelligent and appreciative recipients of them—inasmuch as it acts, in an elevated sphere, on the sound principle of co-opera- tion, uniting its benignly facilitating aids to progress, with the worthy efforts of those " who carry weight in life."
We cannot here stop to give statistics of its practical work- ings, since they are duly stated in its reports.

THE MASSACHUSETTS INSTITUTE OF TECHNOLOGY is remarka- ble for the comprehensiveness, and large scale, of its organiza- tion. It embraces three grand divisions : A *Society of Arts*, in several sections, each devoted to a specific subject of theoretical or practical inquiry ; and working on such a scale,´ as to furnish motive power for use in exhibiting the action of full sized mechanical inventions ; a *Museum of Arts*, analogous to the Paris "Conservatory of Arts and Trades," and a *School of Technology*, in six divisions, as seen in the table, and marked, as it would seem, by a purpose to test the extent to which instruc-

2

tion, in *exact* science, can be effectively given, by lectures, on the basis of sixteen years of age, and an academy preparation, as the minimum of age and training required for admission.

This institution, also has a notable collateral feature, in its system of free evening instruction to intelligent and earnest artizans of both sexes, given in Boston by joint arrangement with the " Lowell Institute."

THE WORCESTER COUNTY INDUSTRIAL INSTITUTE has a quite unique feature, in its unusual proposed provision for the practical study of mechanism. It contemplates nothing less than what might be called a *Laboratory of Mechanism*, to consist of a well appointed machine shop, with power, machines and tools; in which the special student of mechanical engineering can find a counterpart to the Chemical Laboratory of the industrial chemist; the Physical Laboratory of the professional student of physics, (optics, telegraphy, etc.); the Metallurgic Laboratory of the student of mining, and the Mechanical Laboratory (for testing materials, truss combinations, etc.,) of the student of civil engineering.

THE CORNELL UNIVERSITY—to embrace a school of agriculture and the mechanic arts, as a condition of its claim to the share of New York in the national public land grant to the states for the endowment of agricultural colleges therein, contemplated in the law of July, 1862—stands with few or no rivals in the magnitude of its moneyed and landed endowments, The former, including the grand donation of the State Senator, whose name it bears, with the proceeds of the national land grant, amounts, it has been stated, in round numbers, to one and a half millions of dollars; and its grounds are required, by the incorporating act, to contain not less than two hundred acres. The same act allows it to hold an aggregate property not exceeding three millions of dollars.

When it is considered that there is a limit, fast approaching, to the most useful number of such institutions for a given population, having reference, we mean, to the full development of these technological schools, it is most earnestly to be hoped that the organization of this institution will be distinguished by unity, breadth, and comprehensiveness of design, so that, if built up in successive parts, each part shall fall into its fit place as a component of a predetermined organic whole. The opportunity, afforded by its resources, for realizing the ideal of an essentially complete Polytechnic University, is too fair to pass without the most studious and assiduous endeavors to improve it.

In most just, though sad, contrast with the preceding bright array of the crowns of freedom, there appears the shadow of the so called *University of the South, at Sewanee, Tenn.* This proposed institution has met the fate, due to the representative educational head of a frustrated attempt to upbuild the collossal barbarism of a political and social state, on the foundation of a legalized dehumanization of an amiable, docile, and capable race of the fellow men of the members of that state; a state, which, besides being, as respects *humane* civilization, barbarous, was, in the face of the nineteenth century of *christian* civilization, a vast organized practical blasphemy.

Having an offensive dash of haughty sectionalism in its very name, which was, doubtless, significant, this University was, as exhibited in its constitution and statutes, largely pervaded by the sectional spirit of oligarchy and autocracy. It even made a provision, so revolting to a worthy and justly high minded professorial corps, for a counterpart to a plantation overseer, in the person of an officer who was to have very much such disciplinary power over the Faculty!! as the latter should, if at all worthy of their places, have over the immature, or readily misled, youths committed to their (should be) cherishing care. But, in this provision, we only see the form assumed in the field of higher education, by that inextinguishable subtle spirit of *disesteem for labor*, even so elevated as that of the professorial chair—if only it be useful labor—a spirit which is the necessarily blasting accompaniment of a system of bond labor.

In its ambitiously inflated organization, this institution was but a confused collection of no less than *thirty-two* separate schools, so called, some relating only to single, general subjects of study, as Physics; others, to comprehensive departments of professional knowledge, as Law, or Engineering, each properly embracing a circle of such general subjects. We spoke of the above University as having met with a destroying fate. It is reported that its very foundations were carried away piecemeal, as relics, by the armies of *National Unity*, *Broad Humanity*, and *Emancipated Industry*. Let us hope, however, that when, in due time, the spade, the 'loom, the press, and the free school, as secular instruments of free, christianized humanity, shall have done their regenerating work, this institution will reseat itself on its mountain estate of eight thousand acres, as a powerful centre of humane, polite, and industrial culture.

II.

The True Educational Plane, and Methods of Instruction of Polytechnic Schools.

———— ◆•◆•◆ ————

1. EDUCATIONAL PLANE. Systematic education, or the orderly development of the powers of the human mind, by the aggregate of methods and appliances employed in school instruction, exists in four grades, viz:

> *Rudimentary,*
> *Elementary,*
> *General, or "Liberal,"*
> *Technical, or "Professional."*

These grades, or successive stages, are, moreover, natural and not artificial, since each has its peculiar, and strongly marked, defining characteristic. Neglecting, here, their recognized varieties and subdivisions, they may be defined as follows:

1. *Rudimentary Education.*—This is the germ, embracing the alphabet; reading, of merely narrative or declarative sentences, of the simplest kinds, about the commonest things; writing, of detached letters, or their mere elements; singing, by the ear; observation, of common things; arithmetic, of operations on small whole numbers, so small as to be realized in thought.

2. *Elementary Education.*—This initiates the mind into the beginnings of the use of the keys of knowledge. It opens to view, and teaches, Arithmetic, Grammar, Geography, Composition, Domestic and Neighborhood Morals. Indeed, it acquaints

the mind with the elements of the many branches of knowledge, as pursued in Grammar and High Schools, and the
equivalent private schools.

3. *General Education.*—This begins, when the mind has
so far developed as to have an original, free, love for knowledge, and becomes conscious of individual intellectual, artistic,
or moral tastes of its own. This education, it is the *characteristic office and aim of the college to afford.* These institutions
give, to the awakened, eager and active mind, facilities for
gaining a *comprehensive view*, as from a hill top, of the whole
field of knowledge. They also labor to secure for their members such a degree of acquaintance with the various mathematical, physical, philosophical, and classical studies, together
with invigorating practice, by composition and declamation, in
the enlarged use of written and spoken language—such a
degree, we say, of all this, as qualifies the mind, thus " liberally"
trained, to choose which select group of studies it will after-
wards more fully pursue to a practical end.

4. *Professional Education.*—This, when found in the most
favorable condition, is planted in, and grows out of, the well
prepared soil of liberal general culture just described ; or, to
change the figure, it is erected upon that as a broad and substantial basis. Its office and aim is, to give the due, full, and
exact training, necessary for qualifying one for that successful
and honorable professional practice, in which trained and cultivated intelligence is the prime agent in the mere gaining of a
livelihood, but, better, in the life work of making a sensible
contribution to the commonwealth.

By now comparing the professed objects and actual results,
of at least the more well developed, of the institutions named
in the table, with the foregoing principles, we learn, that, at
least in their two or three upper years, they are strictly and
fully *professional schools.* For Civil, Mechanical, Topographical, and Mining Engineering, Physical and Chemical Technology, and Architecture, are not taught in them merely to
discipline the mind, or to qualify one to participate in the

15

intercourse of polite society, though, together with previous general culture, they should richly contribute towards accomplishing these elevated and most desirable objects. These great subjects are taught, principally, as elevated scientific practical *professions*, that is, as means of gaining ample and honorable support, and of ennobling the state, by the application of fruitful principles of science, to the beneficent arts of peace.

Summarily, the end of *College education* is the discipline of the mental faculties, as working forces. That of *Professional education* is the endowment of the already fairly disciplined faculties, with the principles of exact science and applied learning, considered as instruments of higher, productive and physically, socially, and morally, conservative, industry.

Going through this, or any land, with these determining definitions in hand, there would be no difficulty in distinguishing its professional schools, of every name and kind, or however disguised by unfamiliar names, or other irrelevant particulars.

But to present the Polytechnic class of professional schools as in the focus of vision, a distinction must be explained.

Science is *subjective*, relating to man himself, his physical and spiritual constitution; and *objective*, relating to all external nature. In the former, lies the foundation of the ancient professions of Medicine, Law, Divinity, and Polite Literature as a Fine Art. In the latter region of science, lies the foundation for the distinctively modern technological professions of Engineering, Applied Physics and Chemistry and Natural History, and the material fine arts, of Architecture, Music, etc.

Schools, then, alike truly professional, and equal in dignity, as determined by either of *three* decisive tests, viz : The *talent* demanded by them, the *extent and elevation of their courses of study*, or the *magnitude and beneficence of their results*, stand in two distinct groups, appropriately distinguished as *Humanistic*, or *Polytechnic*, according as their chosen scientific field is subjective, or objective ; relating to *Man in himself considered, or to External Nature as able to be richly tributary to man.*

In case of any to whom the previous statements and conclu-
sions of this section are new, and who hesitate about accepting
them till reassured by the argument from competent testimony,
it may be sufficient to refer them to the official publications of
such high and well established institutions as Harvard and Yale
Colleges, or the Massachusetts Institute of Technology and the
Rensselaer Polytechnic Institute. The two former, in that
simple, matter of course way, which is the strongest form of
assertion, as if the question admitted of no dispute, speak of
their scientific departments, as professional, equally with their
other professional departments. The latter `uniformly assume,
as a thing everywhere understood by the well informed, that
their courses are professional ones, in the full sense. And
numerous other scientific institutions, both the detached class,
and those which form professional departments of colleges, do
the same. This question, then, of the grade of Schools of
Technology, may therefore, it is to be hoped, be considered as
finally settled.

2. METHODS OF INSTRUCTION. From a different point of
view than the one here occupied, this topic might justly claim
a full section, or even a separate treatise. But it serves our
present purpose to mention it here but briefly.

The method of instruction in the old professional schools, is
largely that of lectures. Hence, some seem to be ambitious to
have the same method prevail in polytechnic professional
schools also. But we think the connection between the two
things—the grade of the school, and the method of teaching—
is mostly arbitrary, and that the methods of teaching are properly dependent, rather, upon the *nature of the subjects taught.*
Now it is well known, or may be readily understood, that all
knowledge of mathematical subjects must necessarily be exact,
or worthless. Hence, a point lost, or misunderstood, in a mathematical lecture, may occasion hours of discouraging perplexity,
and annoying possibilities of one's entire work in writing up
the lecture being vitiated. Therefore, we would restrict
lecture instruction to descriptive subjects, in which an error

does not vitiate the whole; and to experimental subjects, which address themselves largely to the senses; and to mathematical subjects, only in case of comparatively mature, and considerably proficient, students of them.

Nor do we think that instruction loses anything of freshness and interest—very important elements, most truly—by this method. For, in studying from a text book on exact science, the student has the pleasing certainty that he has a reliable authority to work on, and from; then annotations and reductions of his own, familiar expositions and supplementary notes by the professor, and, in case of Descriptive Geometry, exhibition of curious special cases, and of models, with informal expositions, will, altogether, maintain due interest among those in whom any method would enlist earnest effort.

We only care, now, to add to the above the bare statement of the methods of polytechnic instruction, viz:

Formal lectures.

Familiar expositions, in part conversational.

Interrogations and Black Board demonstrations.

Practical exercises in Geodesy, Astronomy, Physics, Chemistry, Botany, Geology, Graphics, and Mathematical and Mechanical problems, etc.

Excursions, for inspection, sketching, etc.

Systematic Reviews.

Oral and written examinations.

A notice of methods of instruction may, however, embrace a few words about professorial and tutorial functions, and the hours of daily duty of teachers and students.

A professor, properly and distinctively so called, makes some extensive subject a field for continued research, either with a view to enlarging the area of existing knowledge with respect to it, or the bounds of it as actually taught in the place of his chosen labors. He also is the responsible head of his own department of instruction, and gives instruction personally, in the higher subjects of his department, and through assistants in its more elementary portions, taking care to duly superintend the matter and manner of their instructions.

The importance of providing such amount and competency of assistance as will relieve a professor from being merely a tutor, ending the year, so far as advancement of his department is concerned, just as he began it, is clearly recognized by higher educators, and in the practice of liberally managed institutions, since hardly anything conduces more to their vigorous life and growth, than due provision for professorial research, in behalf of increased and remodelled matter, and methods, of instruction.

As to daily labor in polytechnic schools, we believe it true that they are quite generally understood not to be abodes of luxurious ease, or dissipated idleness. Rather they are designed to correspond to the most approved mechanical motors built by their professional graduates, in yielding the largest percentage of useful result in a given time. It is definitely stated, that in the Central School of Arts and Manufactures, at Paris, eight and a half hours in the school, and four more in his room, is the daily standard of student's work ; and similar information is in our possession relative to other European schools. Let us see what an exhibit, for the performance of the human engine, can be made on this basis. Eight hours for sleep, an hour and a half for dinner, and an hour for each of the two other meals, including healthful repose, or light pastimes, makes eleven and a half hours, leaving the twelve hours and a half for work. Now in these hours, mind and body labor conjointly. In some practical exercises, as in a good deal of Laboratory and Drawing Room practice, and in Engineering Field Work, the activity is largely physical, and in the latter case, as well as in out door pursuit of any department of Natural History, is highly pleasant and invigorating. Also in all practical exercises under instruction, and attendance on the more informal expositions of the instructor, there is a subdued play of the kindly social element, which is by no means to be overlooked in its lubricating influence upon the workings of school mechanism. So that the purely mental activity of the twelve and a half hours, reduced to its equivalent of close study, would probably not average half that time, or more than six hours daily, which was about the standard approved by Sir Walter Scott, when at the height, equally, of his health and his success.

There would seem, therefore, to be no difficulty in realizing the preceding programme, tempered too, by a half or whole secular day's absence of prescribed exercises, and the inviolable Sunday privilege of rest, and opportunities for self-adjustment and accumulation of moral power—if life be not clogged with surfeit, like a locomotive choked with a fire-box filled solid with coal dust—if it be not wasted by vice, like the locomotive with inly corroded boiler, that can hold but faint working pressure— if it be not consumed by destroying excitements or stimulants, like the boiler through whose flues, uncovered with water, the fire rages with unnatural heat.

Modern civilization is bound to justify itself by producing a more perfect type of symmetrically developed manhood than before appeared, and the polytechnic school, as a favorite son of that civilization, is bound to exhibit in the sustained activity of its members, a higher percentage of effective work, than any other organization can show.

III.

The Nomenclature, Spirit, Usages and Discipline
of Polytechnic Schools.

—————•+•—————

1. NOMENCLATURE. *a.—General Nomenclature.*—To treat this topic clearly, settled definitions, if possible, must be given to certain educational terms, which are well known to be popularly used in a very loose manner.

First. "*College.*" Turning from the dictionary to an encyclopœdia, for fuller standard information, we find a college, in its primary meaning, to be a union of persons, having "like powers, privileges, and customs, in one office, for a common end." Thus the phrase, "College of the Apostles," is in use to this day, and in the ancient Roman State, trade associations, as of carpenters, bakers, etc., etc., were called colleges.

Again, all through the middle ages, and to the present time, various protective, administrative, judicial, elective, and religious bodies were, and are, called colleges. Thus, there was, perhaps is, the poor men's decent burial *college;* the Russian "*college* of general superintendence," (of benevolent institutions), the "*college* of justice," or supreme court, of Scotland ; the United States *college* of presidential electors, etc.

Lastly, and chiefly, the word "college," in connection with higher education, has a curious history. In that revival of learning, which occurred in the 13th century, celebrated lecturers drew eager crowds of youths to their lecture halls, and special buildings, under proper superintendence, were provided for their meals and lodgings. These were the original colleges, mere endowed students' hotels, both in England and on the continent. These, sooner or later, became transformed into places of instruction, including the lecture rooms within them, and each possessing a faculty of instruction, so that now a "commons," or general eating room, in a college, is the dying relic of what the entire college originally was.

The name of college is seldom applied to professional schools, though Medical Schools, and these only, if we are not mistaken, sometimes call themselves Medical Colleges, also the table in Section I., presents one Polytechnic School called a college. But, in either case, it is not to be inferred, that such schools stand on the same educational plane with true classical colleges, or are conducted on merely college principles.

Our limits forbid the introduction of much interesting matter under this head, which may be found in reports, or papers, on superior education.

Second. "*University.*" This word, like "college," had, originally, no reference to an institution of learning, but only to corporations, who may have preferred this title to that of "college," merely to express the *completeness* of their organization, or the universality with which it embraced all, fitted to belong to it. Thus there were, in ancient Rome, "universities" of tailors, etc.

The word became a term in education, in the 13th century, and did so because it expressed the idea of a corporation. such as was formed by an organized body of teachers. It was always, as now, a term of superior dignity, meaning an institution, or corporation, existing for purposes of higher instruction. There were many of these universities in Europe, in the middle ages, of which the first was at Paris, giving instruction in Law, Medicine, Divinity, and in what was then called the Arts, meaning the literature and meagre theoretical science of the ancients. And, as already described, colleges were nothing more than the hotels of the students at those universities.

Finally, at the present time, the term university is used in various senses, some having no definite meaning. *First.*—The German, or continental, sense, of a school superior to modern colleges—called in Germany, gymnasia—in which any single subject, or department, of general science, can be pursued to any extent desired by the student.

Second.—The general English sense, of corporate institutions, intended for purposes of instruction, and surrounded by colleges, as incorporated and endowed lodging places; but to which the university has quite abandoned the work of instruction. Thus the university is a blank form, and the colleges have advanced from merely, each, giving instruction in some one or two branches, to the rank of competitors, with each other, in giving an entire collegiate course, mostly under tutorial instruction, for an academic degree, or a professional degree, in the old professions. Efforts have been made, however, to reform the English Universities in this respect.

Third.—The new and special English sense, of a senate of eminent scholars, with its boards of examiners, called collectively, the University of London. Students from all the other colleges and universities in England, or its colonies, dissenting or otherwise, can obtain degrees from it, by passing its examinations.

Fourth.—The popular American sense, so far as there is a definite one, tends, perhaps, to associate the term university with those institutions which embrace in their design, or actual operation, a circle of professional schools, successive to the collegiate course as, in part at least, their common foundation. Yet, on the one hand, some institutions, of the highest character, in this country, are merely called colleges; and on the other, some, hardly superior to a New England city high school, style themselves *universities.*

Fifth.—There are, in addition, two American special uses, of the term "university." *First*, as applied to State universities, like that of Michigan, which form, each, the crowning member of a state educative structure, whose foundation is the state common school system. The University of Michigan is a favorable example of these universities, having two parallel collegiate courses, of four years each, one classical, the other largely scientific, and both succeeded by professional courses, in Law, Medicine, Chemical Technology, Civil and Mining Engineering, aided by ample and varied cabinets, etc. *Second.* There is the so called University of the State of New York, giving no instruction, but embracing a board of regents, to whom all the academies, colleges, and professional schools, make annual reports—including some meteorological observations—as a condition for receiving their respective shares of the "literature fund" of the State.

Among and, in part, better than all these numerous, and partly confused senses of the term university, the following might be adopted as a standard one, due to the historic, as well as essential, dignity of the term, viz : A university is an institution for instruction, in which, besides professional instruction in one of the two grand divisions of professional schools, humanistic, or polytechnic, (p. 15) provision should be made for carrying those, who have time, means, and inclination for being students for life, through a course as extended as the existing resources of human knowledge will permit. Also such institutions may properly include a foundation general, or collegiate course, congruous, in each case, with their distinctive professional courses.

Third. "*Academy.*"—This word originally meant only a public park in the city of Athens, where Socrates and his chief pupil, Plato, imparted instruction, in their pagan philosophy, to Athenian youth, assembled in its groves. The disciples of Plato were called Academists, and each, on opening a school of his own, called it an academy.

At present, the term "academy" has three applications. First, to a school, usually private, of about the same grade as any city public high school, and intermediate between the grammar school and the college, as the latter is, between the academy and the professional school. Second, to Government Military and Naval schools. Third, to associations of men, eminent in any one or more departments of general or professional knowledge, or art. These are found in all civilized nations, the most celebrated being the five conjoint academies of France, unitedly composing the Imperial Institute of France.

24

These are the French Academy, the Academy of inscriptions and polite literature, the Academy of sciences, the Academy of fine arts, and the Academy of moral and political science.

Fourth. " *Institute.*"—This, also, is a name of very broad application, meaning anything instituted, i. e. set in place, whether, a custom, or a book, or a school, or association of any grade. Nothing can be inferred from this name, of the grade of a school of learning, or association, adopting it, as these range all the way from boys' boarding schools, up to the unrivalled Institute of France, just mentioned.

Fifth. " *School.*"—This is by far the broadest, or most generic, of all these educational terms, being merely any aggregation of appliances, systematic or not, organized or not, which, intentionally or not, act to develope, either well or ill, the human being. Thus, human life is truly a school. Nature is a school. So, too, particular forms and spheres of life, as street life, workshop life, and business life, are schools. Associated opinion, as general public opinion, or sectarian opinions, are schools, and the adherents to such opinions, are, themselves, collectively called schools. Thus we have schools in politics, in theology, in medicine, in art. Also the term school applies to the whole range of express institutions of instruction, from the humblest primary, to the highest professional one.

More exactly, now, a school is any educational organization, complete in itself, whether existing independently ; or, as a component unit in some more comprehensive organization. Thus, there are medical and other professional schools, separate from any college, and there are like schools attached to colleges as their basis. In the latter case, by reference to catalogues, we shall find, first, the general faculty of the whole institution, considered as a compound unit ; then, separate lesser, but complete, "faculties" of the component professional schools.

With reference, next, to the adoption of " school " as the title of the institutions devoted to the last and crowning stage of systematic education under tuition, that is, to professional education, there is a beautiful ground of its propriety. Stated abstractly, as a general principle, it is this : It is quite beyond the capacity of any sounding title to reflect honor upon, or exhibit the honor of, the highest ideas and objects, so that the latter, being self-sufficient, rejoice in the simplest and homeliest names. " Home " is better, every way, than "paternal mansion ;" the "evening star," than the "nocturnal luminary ;" my "love," than my "most distinguished consideration ;" " teacher," than " professor ;" and " school," than "academy," "institute," or " seminary."

This really familiar principle is very generally acted on, in naming professional Institutions, which are almost invariably called schools, both separately and collectively, as Law Schools, Scientific Schools, Theological Schools, etc. The name of school is adopted then, although the simplest, yet as really the highest, because, as above shown, the most generic. The descriptive epithet added, as Polytechnic School, marks both its sphere and grade. This, however, when but a single professional course is given. Each course leading to a degree, demands its special school, and the term Institute, is especially recommended, by frequent continental European practice, as the general title of the organization.

Sixth.—Without making separate heads for the following, a " department," as distinguished from a " school," and as a branch of a comprehensive institution, might be defined as not being subject to a special faculty, complete in itself, included within the general faculty, as before described in defining a school, though it must be confessed, that this definition has exceptions in actual usage. In Germany, when " school " is the *general* name, " departments " are often called " sections." Lastly, " seminary " is not the name of a different kind of institution from those bearing any of the preceding names, but merely a different name for the same thing, a name based on the idea of a school as a place for the dissemination, or seed sowing, of knowledge. Divinity schools, especially, for instance, style themselves indifferently, "schools," " institutes," " seminaries," or " departments."

4

b.—Professorship Nomenclature.—The Chief of internal administration in higher institutions is variously styled, President, Chancellor, Rector, Provost, Director, etc. The last term is appropriate to polytechnic schools, as conformed to continental usage, and as in accordance with the desirable features of essential unity of administration, and an executive organization of chief and associates, analogous to that of a civil chief and his cabinet, or a state governor and council—the chief, in all such cases, having due authority to act singly, in emergencies demanding power and promptitude. But we had more particularly in mind, that very important feature of true department nomenclature, which duly expresses the fact that *each of the scientific professions has large component parts, each forming matter for a full professorship.* Thus, Civil Engineering embraces, as necessary and fundamental to it, Mathematics, Physics, Analytical Mechanics, Geodesy, and Descriptive Geometry, or the Science of Form, with its applications. Now when the separate chairs in a Divinity school, a Law School, or a Medical School, can be consolidated in one; or, when one man can give duly elevated and extended courses of instruction in the five foregoing departments of knowledge, then, and not before, will the phrase "professor of civil engineering" and the enumeration of "civil engineering," as a simple element of a programme of study, co-ordinate with other single studies, as History, Geology, Mechanics, Drawing, etc., cease to be absolutely ridiculous. This assertion, is, of course, no intended reflection upon those who act under such a nomenclature, since they find it ready made for them, and, very likely, tolerable only as a provisional concession to popular misapprehension of the real constituent parts of engineering science.

According to the misapprehension alluded to, civil engineering is about equivalent to geodesy, which is only one of its subordinate components. For the end of geodesy, relative to engineering, is the *instrumental determination of field data,* as a basis for the proper *designing of works,* which last requires an extended knowledge of Mathematics, Technical Physics, (strength of materials, etc.,) and Mechanics; and, then, the *intelligible representation of works,* whatever their complexity, and in all their details,' by an application of the principles of Descriptive Geometry. Hence, in no continental polytechnic programme, that we have yet heard of, can be found any such anomalous expression as "professorship of civil engineering," or any analogous nomenclature.

c.—Class Nomenclature.—Turning next, for a moment, to class nomenclature, we find the numerical system (1st, 2nd, etc., classes) in general use in all lower schools. In colleges, the titles "Freshmen," "Sophomores," "Juniors," and "Seniors," are doubtless unalterable, and well enough so. In some professional schools, classes are designated in partial repetition of the college nomenclature, as "Junior," "Middle," and "Senior," in *three* year courses, or Junior and Senior in *two* year courses, such as are usual in Law and Medical Schools. In others, the mere terms "First year," "Second year," etc., indicate the classes.

In the case of professional schools, having a four years' course, as in two of the polytechnic schools named in the Table, (p. 6), there are manifest objections to a mere repetition of the college nomenclature; since the entering member of any professional school, whatever his previous studies may have been, *stands in a scholastic position* four years in advance of the college, "freshman," and probably does not propose to become, or be regarded as, a freshman a second time, after such an interval. Assuming, then, that the polytechnic variety of professional schools may reasonably have some distinguishing badge, in its class nomenclature, there is reserved for these schools the alphabetical system, adopted by the Rensselaer Polytechnic Institute, also by the Cooper Union (p. 9), for the classes in the five year course of its night school. Only, in the former case, the badge is one of *total* distinction, the classes, being styled "Divisions"— "Division A" (the highest), etc.; while in the Cooper Union, a badge of union with the entire fraternity of educational institutions, together with a duly distinctive nomenclature, is found in the retention of the universally employed word "class"—its classes, above mentioned, being Class E (the lowest), etc.; a very good system, we think, and worthy of general adoption by Technical Schools.

II. SPIRIT.—Passing to the *Spirit of Polytechnic Schools*, it should, in common with that of other professional schools, above all things, not be in any degree a weight upon the neck of the local civilization where it exists, but itself a centre of refinement, no less in its grounds and other material appointments, than in the life of all its members, and in that of its officers. The fundamental social, and moral, qualification — no less important than scholastic ones — for membership in a professional school, as such, is, possession of both *ability and disposition, to act steadfastly in the spirit of a man* — of a *young* man, by all means, but still of a man — ready to be governed by the laws of the land, and by the equally inviolable, though unwritten, laws of social propriety, and of honorable professional life.

Again, in colleges, the unwilling attendance, perhaps, of some, and the absence of any definite high aims on the part of others, and the varied ultimate aims of most, tend to disunite their members, and the existence of secret societies tends, one would suppose, still further to narrow and hedge in a spirit of broad fraternity. But in a professional school, the *unity* of *aim* of all its members, at least of all who contemplate taking the same degree, is a natural basis for that comprehensive unity of feeling, and sentiment of substantial equality, which would render all class jealousies and disaffections impossible, which would make each member regard each, as, *primarily*, a member of the institution as a whole, — *secondarily*, as a member of a particular " school," or class, in it.

Last, but by no means least, every member of every polytechnic, or other, professional school, should pursue his work with free ardor, in the spirit of *voluntary and interested research ;* and not in that of reluctant fulfilment of unwelcome prescribed tasks. This radical element of the professional student's spirit is also most unequivocally demanded by the primary facts of his position. For every candidate for a profession is supposed to have freely and devotedly chosen it ; and this choice involves in it an equally hearty choice of all the labors, and parts of the course of training, necessary for honorable and promising entrance upon that profession. To this end, effective and permanently reliable command of professional knowledge, considered as indispensable to *real and permanent success* in life, will be his absorbing aim. He will therefore never be satisfied with such merely *provisional* knowledge as will serve only the shallow and aimless purpose of a mere technical " passing " of an examination ; while he can but despise all knavish shifts, and aids to the mere *form* of success, without the *reality*, as mean in themselves, and as too pitifully short-sighted, in view of the exacting demands of a professional career. So reasonable is all this, that it would seem, and is, doubtless, generally true, that nothing more than an occasional suggestion — true, earnest, and friendly — could be necessary to hold even a moderately right thinking and well meaning young man steadfast in obedience to it.

III. USAGES.—Out of the proper *spirit* of professional schools, some of whose elements have just been indicated, there will grow a spontaneous rejection of certain inferior and ignoble usages—native in lower schools—and of the sometimes absurd tyranny of class majorities, whenever, for example, it acts, as it sometimes seeks to in lower institutions, to interfere with the inalienable right of each individual student to enjoy and improve every privilege and opportunity offered by the institution which he attends — things which are acknowledged as blemishes, if not as serious evils, in those lower institutions, and in the earlier stages of student life. And, so far as new usages are instituted, they will be made to harmonize with *professional* student life, as the highest and closing stage of that life.

These lower usages, and customs, will be, and usually are, exotics, impossible to naturalize in the soil of any professional school, which is true to itself; and even the best designed secret society should hardly claim recognition as an active organization in such schools, in competition with the other broader, higher, and worthier grounds of fraternity, which have been shown to be afforded by professional student life. Indeed we believe it to be true that secret societies rarely, or never, maintain an active organization in professional schools, subsequent to college courses.

In this connection, however, a much more interesting and important question arises. The legitimate objects and doings of voluntary associations for mutual improvement, if indeed any such should exist in professional schools, presents itself as a subject not without difficulty. As every one knows, nearly, or quite every college possesses one or more large and flourishing literary societies. Their existence is readily justified by the facts that the characteristic office of the college is to develope the mental faculties, and that these faculties are rapidly developed by voluntary painstaking exercise, in view of criticism by quick and watchful competitors.

But the office of the professional school is quite different. It presupposes faculties already fairly developed, and although it does, incidentally, expand, strengthen, and polish them still further, yet this is not its primary aim. For its aim, as before shown, is, to store the capable mind with fruitful truths, that is, with *principles*, and to initiate the eye and hand in the elements of material professional practice, all with a view to a productive application of these principles and scientific physical accomplishments, in subsequent professional life.

Now the determining question is this: Can a *professional* student secure accurate scientific information—which, by its nature, must be exact, or worthless—and practical scientific skill, more rapidly and effectually than by *devoting all his energies to the most faultless possible preparation of all his lessons, and execution of his practical exercises, under thorough professorial direction and supervision ?* The usual practice of professional schools, so far as we are informed, replies in the negative. We are not aware of voluntary associations in professional schools, supplementary to the declared objects of those schools, that is, analogous to college literary societies. Besides, as above shown, the entire course, itself, of a professional school, is supposed, by the very position and proper motives of its members, to be entered upon and pursued, in the free spirit of *voluntary* and *interested* research.

Still, in the polytechnic division of professional schools, we think there is a legitimate, though duly limited, field for the occupancy of voluntary scientific student associations.

First.—They may be made the occasion for the interchange of valuable results of study and investigation, *provided that every member of them is qualified to contribute something, and pledged to do it*, so that all may share the discoveries of each, and thus add to that permanent fund of information which it is a primary object to acquire. The results alluded to may be elegant *mathematical reductions;* lucid *supplementary notes* to obscure passages in text books ; *original solutions* of problems, and *discussions* of their special cases ; contributions of *industrial drawings*—so much more stimulative to student ambition than engravings, or copies made by an instructor—or *models* and *cabinet specimens*, such as can be made or collected in vacations, etc.

Second.—A second general object, in apparently entire harmony with the main objects of the school, would be the collection, through regular correspondence with graduates, and others belonging to the professions taught in the school, of copies of *professional reports*, prepared by those persons ; also the exchange of the various *regular, or occasional, official issues* of similar professional schools, and the collection of valuable pamphlets, etc., bearing on professional education.

Such a society would not exist for purposes of debate, nor would it probably be well, save in case of a very large institution, perhaps embracing a resident graduate staff of high talent, or in conjunction with several other like institutions, collectively sufficient to afford, at all times, an undergraduate staff of high merit, to maintain a periodical publication, inasmuch as a worthy one would otherwise be apt to abstract too much time from devotion to the student's really best interests—already pointed out—as a professional student. The society would be whatever its name, or organization, substantially a " *Society of inquiry*," analogous, in the scientific field, to " Societies of inquiry," in other departments of research.

IV. Discipline.—The actual *Grade*, correspondent *Spirit*, and consequent legitimate *Usages* of polytechnic and other professional schools, being substantially as thus far described, the question of discipline in them is narrowed down to the smallest limits, barely entitled to recognition as a proper question. Every member of such a school, having made free choice of a high profession, cannot but be imaged in thought as diligently devoted to the means of fulfilling his choice, under the kindly guidance of his teachers, whom he will be necessarily incapable of regarding otherwise than as, only and always, co-operating with him, to secure most fully the end he desires, and, thereby, incidentally, to promote the best honor and welfare of the school, with which both parties are identified in spirit. Where, it may well be asked, is there room for the idea of discipline in such a picture?

But let us proceed to search into the elements of this topic. For though it may cover ground very familiar to many, conversant with classical colleges, and the variety of professional schools, which have been called humanistic, yet to the newer community of scientific general intelligence, and eager interest in general and technical scientific education, such a re-discussion may not be untimely.

The administrative affairs of the higher schools of learning resolve themselves, then, into two main divisions : their *external* or *material affairs*, and their *internal* or *immediately educational* ones.

These two classes of interests, being quite different, though intimately connected, are, in common practice, as by natural propriety they should be, committed to two distinct, yet, though in separate spheres, really co-operative bodies, viz. : to a Board of Regents, Overseers, or Trustees, and to a Faculty, embracing, or not, the entire professorial corps, according to its numbers, and other obvious considerations.

A Trustee is one to whom is committed the execution of a trust ; and, in case of permanent institutions, as those of learning, this execution includes, as cardinal elements, the *establishment, maintenance,* and, if credit is to be given to the founder, as a growing, progressive man — provision for the *growth* of the institution.

But, by expanding this statement somewhat, we have the following view :

1. *The external affairs,* embrace these principal points:

1. The holding of the course of the institution true to the general plan designed by its founders ; so that, for example, no medical school could be transformed by its faculty into a theological one ; or a classical college, into an academy of music.

2. The construction and equipment of fit and necessary buildings, located on suitable and sufficient grounds ; the buildings to be designed, as far as desirable, by their professorial occupants, or with their approval and supervision.

3. The provision of adequate compensation for professorial work demanded, according to a justly recognized value of the same.

4. The appointment of officers of instruction, which, to best promote desired success, should be in accordance with nomination, recommendation, or known approval of other such officers, if already existing in the institution.

5. The holding of an existing faculty responsible, in behalf of material interests, for the successful working of the institution, unavoidable external hindrances excepted, under a system of instruction and government to be devised and administered by the faculty ; and expecting them singly, or severally, to give place to more competent successors, if their department, or general systems and administrations, respectively, manifestly fail of success, owing to inherent imperfections.

6. The establishment of appropriate regulations for preserving the buildings and other property of the institution, and for the management of its funds ; also in some cases a certain extent of active participation in forming outlines of a system of rules of internal government ; especially for academies, and for institutions of the collegiate type, particularly for State Universities, like that of Michigan, for example, which, *being creations of the people,* may reasonably be regulated, in a general way,

5

34

by agents chosen by the people ; as is done in the case referred
to, but with an important qualification, soon to be noted.

These high and honorable functions are committed, as before
stated, to a Board of officers, chosen, in part, for their possession of such liberal culture, and enlarged views, as would
make them readily sympathetic and co-operative with an earnest Faculty, in appreciating, and laboring to meet, the claims
and wants of an institution ; and in part for their possession of
business capacity and energy to secure, in conjunction with
Faculty efforts, due pecuniary response from wealthy liberality,
to these claims and wants.

2. *The internal affairs* of superior institutions, are ranged
under these two principal heads :
1. *Instruction.*
2. *Government.*

The department of *instruction,* in a general sense, includes
the designing of a comprehensive and symmetrical curriculum,
in harmony with the declared objects of the institution, and of
a practicable daily working programme, as a means of realizing
the proposed curriculum, as well as the actual work of class
instruction.

The department of *government,* embraces the equitable and
charitable, while efficient, enforcement of such written rules as
are found expedient, for those institutions which are fit subjects
for government under the system of written rules, viz : academies, and, in part, colleges. It also embraces the strict holding of professional students responsible for violations of the
obvious proprieties of their position, without rules of general
moral or social conduct, either to *instruct* or to *constrain ;* these
being the legitimate functions of rules. For the whole theory
of a professional school supposes that every member of it is, as
before stated, both able and willing, by virtue of the very
nature of his position, to do his duty as a student, man and gen-
tleman. If he is not thus able, owing to social or moral back-
wardness, nor willing, owing to obliquities of moral purpose,
he is simply out of his proper position. Accordingly, with the

clearly pronounced moral character, properly correspondent with the general maturity of mind and character naturally belonging to membership in any professional school, every member either is, or is not, entitled to his position. If he is fully, or nearly, so entitled, or is readily accessible to influences tending to make him perfectly so, he should be retained. If he is not, he should be promptly *exscinded*, we would say, not "expelled," as *appended to*, but in no true sense *of*, the proper membership of the institution. The professional school is no field, we think, for the exercise of that tentative, or expectant, method of discipline, which consists in a long drawn gradation of penalties, embracing college rustications, etc., etc., etc. Indeed, it is not such a field, in prevalent practice. But of this somewhat further, in the next section, as it cannot be discussed just here, without too much complication of the topic immediately in hand.

Now to whom are these internal affairs legitimately committed? To the faculty, as supreme, acting under the abundant regulative agency of a general, but high, responsibility, already explained, for the success of its administration. This position is no less supported by sound reason than by prevalent usage.

First, in reference to *instruction*. A curriculum must be made, first, to accord with the declared objects of the institution adopting it. Then, as the time demanded for completing the course of study required by it, also material alterations in the length of the course, may decidedly affect the financial prosperity of an institution, through effects upon the attendance which it can command, these points are matters for mutual conference and agreement between the officers of external and internal government. But beyond these general preliminaries, the control of the officers of instruction, over the arrangement of studies, and methods of teaching, is probably nowhere questioned.

Second, in reference to *government*, several *rational* grounds for supreme faculty control present themselves.

1. If at all competent to their other duties, as teachers in a professional school, would not men intellectually capable of

giving the elevated instruction expected of them, also know what and how much of student propriety to ask, and how to secure it?

2. Principals of academies, may, in many cases, regard their positions as provisional, while seeking some other, as a permanent one; but professors in superior institutions, usually contemplate their positions as permanent, unless called to better ones, and enter into their duties as more or less a labor of love. They identify their own reputations with that of their chosen institution, and thus having *every motive* to study and promote its welfare, and *no motive* to defeat that welfare, they are under no dangerous temptation to do deliberate injustice to any one under their care. Besides,

3. Which is worthy of separate mention, they act, according to their legitimate form of responsibility above mentioned, knowing that they justly forfeit their places, if a system of their own free devising, and externally unhindered administration, manifestly fails of success.

4. And not least, how could those who, week after week, and month after month, come in daily intimate contact with the members of an institution, but be infinitely better qualified to deal justly with offences, than those who rarely, or never, meet with those members?

Testimony also is clear in support of our position. Two representative specimens will here be introduced, since some are so constituted as to be better satisfied with the argument from experience and testimony, than with a purely rational one.

1. From the "Seventy-sixth annual Report of Regents of the University of the State of New York," 1863. After noticing that some academies had lapsed into partial inefficiency, and attributing it immediately to want of the exercise of trustee supervisory care over their internal affairs, needed, perhaps, for the reason just now explained, they proceed thus; "*The faculties of colleges are necessarily intrusted with their internal administration.* (The italics are ours.) Composed of gentlemen, of experience and ability, who, in most instances, have chosen their profession as the employment of life, their character being that

the institution with which they are connected, they have every
motive to faithful and earnest duty."

And it only needs to be added : If this be true for colleges,
how much more for professional schools, of every kind, as
belonging to the next succeeding educational stage.

2. Extract from the constitution and laws of one of our largest and most successful universities :

For the General Department.—" The immediate government
of the department shall be vested in the faculty, and it shall be
their duty to direct and instruct the students in the several
branches of learning taught in the department, [to encourage
them in the acquisition of knowledge and the practice of virtue,
to counsel and warn the offending, and faithfully and impartially
to administer the law established by the Regents,"] the last
phrase being in accordance with the fact that the institution is
a creation of the people of a state, and therefore under a general
supervision, by agents periodically elected by the people.

The whole paragraph, it may be added, is happily expressive
of what every worthy professor voluntarily and gladly does.

For the Professional Department, taking the medical school as
an example. " The immediate government of this department
shall be vested in the faculty, whose duty it shall be to instruct
the students in the several branches of learning, taught in the
department." This is all, and in addition to the testimony to
the lodgment of control over *internal* affairs solely with the
Faculty, how significant the omissions, how strong the assertion, by implication, that every member of a *professional* school
is responsible for being a self-governing man, in spirit ; to stand
in, or fall from, his position, according to his conformity to that
standard. Indeed, in the report of the Regents, just before
referred to, the almost stereotyped phrase in the separate
reports of the numerous professional schools, is, " No rules of
discipline have been adopted. General propriety and decorum
are required."

Once more, an instructive citation, from the same source,
merely to show what, and how much, is meant by the vesting
of the *internal government* of all departments in the faculty

alone. "The presenting of petitions, or other papers, to the Board of Regents, in regard to the government of the University ; etc. ; etc., are regarded as disorderly ; and any student who engages in such practices may be dismissed from the University *by the faculty* (italics our own) of the department to which he belongs."

In view now of all this extended re-discussion of ground, embracing well established principles and usages, familiar to many higher educators, no anomaly could be more evidently unseemly than would be the extension of the college system of rules, with pains and penalties annexed, over the superior domain of professional student life, unless it should be such an extreme misapprehension of the grade of the polytechnic class of professional schools—as level with that of other professional schools—as would lead to the sinking of them even below colleges, to the plane of such academies as might seem to be in need of an active trustee administration of their *internal affairs*, as well of their *external* ones.

It is only necessary to add, in conclusion of the remarks under this head, first, that they are not a plea for what is not, but ought to be, but are the result of inquiry as to the natural grounds of the usages already generally established, by common consent, as right and proper ; and second, that nothing now said militates against the existence of rules for the proper use and care of special rooms, and conduct of special exercises, as Laboratories, Observatories, etc., Field Exercises, etc.

IV.

𝔇iscrepancy between the 𝔜deal and 𝔞ctual 𝔓osition of 𝔓olytechnic 𝔖chools, and their consequent radical 𝔚ant.

———————◆———————

There is no motive for concealing the fact that the preceding views are, in part, ideal, because, in a few of the most developed cases, the actual so nearly approaches the ideal in many substantial particulars, or can easily be made to do so, in these, and other, cases.

In reference to instruction, the great want of polytechnic professional schools, is a class of preceding institutions, bearing the same relation to them, that a classical college does, for example, to a theological school. This want is, however, not totally unsupplied. For, first, Norwich University, Vt., Michigan University, Union College, the University of New York, Brown University, and some other institutions, expressly set forth two parallel courses of general training and liberal culture, the one classical, the other substituting the French and German languages of living and fruitful science, physical science itself, and modern history, for ancient history, and the dead languages of still more dead gods, and their corrupt intrigues. Other colleges, as Harvard and Yale, partly accomplish the same thing by a more or less liberal provision of elective studies, embracing mathematics, physics, natural science, modern languages, and history.

Every distinguished and high-minded professional man earnestly desires, by his love for his profession, that every one

40

entering it should possess a previously acquired liberal education ; either a collegiate one, or *the nearest attainable substantial equivalent for it* that the still incompletely organized and classified educational instrumentalities of the country allow, in preparation for that profession. But, as is well known, there is a want of adaptation, on the one hand, of collegiate culture to the wants of all the different professional schools, and a readiness in the community, on the other—happily decreasing it may be hoped—to accept boldly self-asserting superficiality. Wherefore, it comes to pass, that, in looking through the catalogues of professional schools, we find it not insisted on, as a condition for admission, that their members shall be college graduates. and but few of them are. A few scattering statistics will sufficiently illustrate this point, as seen in the following

PARTIAL TABLE

OF MEMBERS' OF PROFESSIONAL SCHOOLS, HAVING COLLEGE DEGREES.'

	LAW.	MEDICAL.	SCIENTIFIC (Technical.)
Dartmouth, 1864,		8 of 47	0 of 37
" 1865,			0 of 48
University of Michigan, 1858,		5 of 137	7 of 36
" " 1865,	48 of 260	25 of 414	7 of 29
Harvard College, 1851-2,	71 of 104	34 of 116	17 of 69
" " 1861-2,	53 of 103	45 of 206	13 of 57
" " 1863-4,	75 of 123	50 of 167	15 of 75
Yale College, 1863-4,	12 of 31	10 of 45	7 of 57
Union College, 1860,			2 of 46
" " 1865,			5 of 40
Columbia College, 1864-5,	88 of 158		3 of 43
Philadelphia Polytechnic College, 1864,			2 of 136
Rensselaer Polytechnic Institute, 1860,			3 of 75
" " " 1865,			10 of 152

The above results show that all professional schools stand in an attitude of compromise. While their most earnest friends would like to see every member of them possessed of a "degree," representative of a previous "liberal" or general training, they must accept the nearest attainable equivalent for it. Considering, now, at what a disadvantage the scientific technical schools are placed, in the scarcity of collegiate institutions giving a previous general culture, suited to their wants, the fair proportion of collegiate graduates among their members is surprising, and gratifying. In connection, too, with the undoubted fact that many others of those members have, by diligence, and pursuit of extra studies in the best academies and high schools, obtained the substantial equivalent of a college education, the above proportion of graduates is a new vindication of the claim of these technical schools to full recognition as professional schools.

Definite statistics, in respect to the nature and extent of the previous studies of members of the "Scientific Schools," are, of course, not very readily obtainable.

The following view exhibits the results of inquiries, for three times of admission to the Rensselaer Polytechnic Institute. Out of 132 men, of whom inquiry was made, the figures below show how many had studied, more or less, the subjects against which the figures stand :

Botany,	32	History,		75
Chemistry,	78	Composition and Rhetoric,		92
Geology,	26	Mental Philosophy,		24
Physical Geography,	48	Moral Philosophy,		28
Natural Philosophy,	101	Greek,		25
Physiology,	47	Latin,		65
Astronomy (Popular),	49	French,		65
Music, Vocal or Instrumental,	49	Other Languages,		25
Free Drawing,	50			

This result gives nearly *seven* subjects, on an average, to each man, besides the fundamental subjects for admission, viz :

Arithmetic,	*Elocution,*
Algebra,	*Penmanship,*
Geometry,	*General Grammar,*
Geography,	*Orthography,*

and besides something done, perhaps, in

Zoology, etc.,	*Geometrical Drawing,*
Logic,	*Book Keeping,*
Political Economy,	*Trigonometry,*
	Surveying,

subjects not embraced in the inquiry, though they very pertinently might have been.

6

Much might be gained to the cause of sound and advanced scientific professional scholarship, by the general adoption of the *Elements of Physics* (Natural Philosophy), of *Trigonometry*, of *French*, and of *Geometrical Drawing*, as requirements for admission to Polytechnic Schools, in addition to the eight subjects above mentioned.

A second method, by which the polytechnic institutions are supplied with due preparatory courses, is by carrying backward their own courses of study behind the point at which they are wholly or strictly professional. As may be seen by reference to numerous catalogues, two years, only, occasionally three, is the usual length of Law and Medical courses of instruction, the commonly required three years' residence with an approved practitioner, in the latter case, being offset by the subordinate positions generally occupied by young engineers, etc., for an equal time. But, by reference to the Table in Section I, we see that the scientific school courses are frequently of three, and sometimes four, years' duration. Now, in several of these institutions, the earlier portions of these extended courses, embracing as they mainly do, subjects which every one, aiming at a high standard of "liberal" scientific culture, should be acquainted with, are expressly placed within the sphere of collegiate, or *general*, disciplinary culture. Thus, at the *Philadelphia Polytechnic College*, there is a separately entitled general "scientific course" of one year, disclaimed as professional, surrounded by a circle of six professional courses of two full years each. In the *Massachusetts Institute of Technology*, the first two years study, while evidently designed to correspond to a very elevated standard of what general scientific training should be, is only assigned to the sphere of such training, while the several parallel courses of the last two years are designated as strictly professional. And once more, in the *Rensselaer Polytechnic Institute*, whose course is one of four years, the studies are, from the beginning of "Division D," narrowly, but increasingly, and at last almost purely, professional; and, correlatively, at first widely, but decreasingly general; or of a kind necessary to be understood by persons desiring only a liberal disciplinary education.

From the results thus far indicated in this section, two important inferences, and a concluding reflection, arise.

First.—The Officers, Members, Alumni especially, and Friends generally, of technical schools, have a mission to perform, in elevating them to an *unobscured*, and *undisputed*, level of rank, with the universally acknowledged professional schools of other kinds. This mission embraces such particulars as the following: 1. As college graduates, other things being the same, naturally make the most appreciative and well qualified members of professional schools, every effort should be made to increase the number of those colleges which afford *scientific general courses*, of not less than three years' duration, as the legitimate forerunners of *scientific technical courses*. To expedite this desirable movement, academies also—for in them the work must begin — should divide their upper classes into sections, the members of one of which should be put in special training for a *scientific* college course, while the members of the other would be preparing for the parallel *classical* course. 2. That the professional rank of the technical schools should be unobscured, the more fully developed among them, so far as they desire to do their own preparatory training, might well resolve themselves into a distinctly pronounced *two-fold general organization*, the first department of which should be of a collegiate character, and adapted to the earlier wants of youth seeking a finished scientific education ; the second department embracing any proposed number of strictly professional schools, managed exclusively as such, in respect to matter of instruction, and tone of administration. 3. The establishment of *resident graduate, or true university, courses*, according to the standard, named on p. 23, for the benefit of those who have means, and desire to pursue particular subjects to an unusual extent ; also, efforts to secure, at all times, at least a few students in such courses, who would also peculiarly benefit, both themselves, and the institution, by becoming assistant instructors in it.

4. The general adoption, in full, of the three fundamental tests of student proficiency, viz :

a.—The *daily* recitation, or interrogation, upon assigned lessons, or performance of assigned exercises, and solutions of new problems, as the distinctive test of *regular daily fidelity* to duty, and growing *command of principles;* first, in advance; second, in review.

b.—The *oral* session examination, or test of power to *retain* matter once learned.

c.—The *written* examination (upon new applications of general principles) the test of *retained available command* over one's knowledge, for purposes of varied practical application.

The examinations, should, moreover, to possess the greatest value, cover three different periods,—first, each *term* as a whole, second each *year* as a whole, third the *total course,* as a whole, so that the graduate could, most truthfully, as by the law of public morality bound, be represented as possessed, *at graduation,* of at least a fair available knowledge of the entire course of study pursued by him.

The efficient maintenance of these three tests, and legitimate *external* stimulants, on the one hand, and natural adaptation, as the natural *internal* stimulant, on the other, might doubtless be relied on to secure results, permanent and solid, if not brilliant, and such as would demonstrate the impertinence of every artificial stimulant, such as prizes, etc., etc.

5. With the adoption of such essential measures as the above, the merely formal representative, but very desirable, ones, of increased age, and scholastic requirements for admission to technical schools (see p. 42), would fall into place as matters of course. They are worthy of separate mention, however, since their adoption would doubtless react, especially in conjunction with the fourth particular just named, to secure the desired movement in respect to the first three of the above fundamental measures. It is our conviction that the best rule for settling the somewhat arbitrary point of age for admission, would be, to subtract the total length of the course from the age of twenty-one years, as the minimum for *professional* graduation.

Second.—In view of the more or less mixed character of most existing Technical Schools, as now explained, the grand question arises, shall their *governmental administration* accord with the provisional, abnormal, and subordinate general, or collegiate, character found in their earlier stages, or with their *permanent, normal,* and *more and more prevailing character,* as purely professional schools? With the latter character, by all means, we most heartily say, after much experience, with many a company of efficiently self-governing young men. If a single qualification is to be made, as à provisional concession to the mixed character of our Technical Schools as at present found, it would be in favor of the adoption of the single rule requiring regularity of attendance, and responsibility for preparation, since, when these points are secured, nearly everything is secured, so true is it that idleness is the open door to every vicious folly. For all the rest, uniform conformity, without rules, to the standard implied in previous statements, is to be tacitly demanded, and practically enforced, quietly, and as matter of course. But while the inviolable honor of a professional school demands this plain speaking, it should be regarded, *first,* as no less the voice of all its members, than of its Faculty ; and *second,* as in no way inconsistent with that sacred regard for human nature in the stage of young manhood, which would, by every kindly means, forestall all need of discipline.

Few are so strongly self-centred, through possession of that controlling personality, which consists of a vigorous will, guided by enlightened reason, as to be the same, in character and conduct, under the strain of greatly varied surroundings ; as to be free from the sway of the principle that men will often be to a great extent what you, by your manner of dealing with them, practically declare them to be. Wherefore, if a professional school is operated on college or academy principles, i. e., under a code of formal rules — too often embracing petty provisions, or commanding, and enforcing by an espionage, humiliating to all concerned, those higher duties, performance of which must be free, or worthless — the characteristic blemishes found in the weaker and frivolous elements of college and academy life, will find their familiar " *habitat,*" and spring up with the certainty of fate. But, conduct the almost completely professional school in the interest of its own best aspirations to be undisguisedly and undisputedly such, and there is abundant and bright evidence to show, that, even with its youngest members, regard for its honor and dignity, as well as for the home whose wish is law, will maintain all needful supremacy over the natural impulses of earlier young manhood. Why then repress this rising, and easily cultivated, spirit of healthy manliness, and professional honor; and, *for no equivalent good secured*, postpone the full attainment of the acknowledged rank of professional school for, and of, young *men*?

But the most complete and decisive justification of the policy, here advocated, lies, it seems to us, in the obvious propriety, if not positive obligation, of making the closing stage of a young man's student life correspond, in its prominent features, with the closely subsequent practical life, in which he *must* stand, or fall, according to the amount of his own knowledge, and power to use it, and according to his self-governing power. Is it justice, we ask, to the unalterable constitution of human nature, to plunge it at once from a system of floats, and guide-ropes, in a shallow tank, into deep and troubled water, where the powers of a *practised* swimmer are required? Are not educators for professional life *bound* to afford, by a system of administration which demands substantially self-governing manliness, a little experimental, and last, school circle of practical life, preliminary to the world's great circle of real life? Should not the discipline of the professional school, as the closing one, be stimulative of interest and alacrity in the good work of self-discipline and early self-government, instead of listless or murmuring obedience to ignoble external restraints? Why should the character of the final system of control over student life be based on the conduct of the meanest few, who have no claim to their position, rather than on that of the honorable and self-regulating many? In other words, *why should it be based on a few mean facts, rather than on many goodly ones*, so as to present to all right endeavor the pledge of the best recognition, viz., recognition of its right to real freedom. And here we add, that every member of every kind of professional school, who would

47

see, and be animated by, what is practicable in self-government, under rules, courts, and procedures, of their own devising, among such students, and practicable in elevated and refined associate life, would do well to ponder the account of a celebrated Swiss school, described in the article, "Student life at Hofwyl," in the Atlantic Monthly for May, 1865, an article which, it is to be wished, might be separately printed for wide circulation as an educational tract.

As this article may not be accessible to all, we cannot exclude an intimation of the character of the system described in it. According to this system, the primary disciplinary power of a superior institution should be its students themselves, acting through an organized and dignified tribunal with regular rules of procedure, and acting in behalf of a high-toned student civilization. The decisions of this tribunal, in reference to offenders against the true honor of the institution, were to be subject to revision, or absolute veto, by the Faculty. The practical effect of this feature was, however, to stimulate the students, strongly, to weigh and consider their decisions so dispassionately and carefully, as, if possible, never to have them vetoed; and even modified, as seldom and as slightly as possible.

Under such a system, the well-being of school buildings, and the absolute immunity of its furniture from all needless defacement, could never be more complete than when committed to the voluntarily responsible charge of the students; while nothing could so restrain idleness, drunkenness, or offences against neighborhood peace, or property, or disorderly concomitants of out of door exercises, or excursions, so effectually as wholesome sense of strict accountability to the high-toned collective sentiment of one's peers, enforced, through the orderly action of a tribunal of those peers.

It is also but justice further to add, in finally dismissing this topic, that the writer, himself, attended, for two years, a private free school* of high order, in which no code, if it existed, was ever posted or heard of, and in which the grounds were

---

* In Newburyport, Mass.

laid out, and well kept, by the pupils, and the building was treated as a home by them, and all the relations of teachers and pupils were those of a polite company, bound together, and to duty, by unwritten laws of social decorum and kindness. But it should be added, in partial explanation of this elevated character of student life, that this school embraced pupils of both sexes, who associated freely, under the fewest guiding restraints, not only in daily classes, but in musical and horticultural associations, and in editorial and anniversary managing committees, all of which were active organizations. Rational faith, in young humanity thus put on a fair footing, here had its perfect reward, in the absence, nay more, the practically impossible occurrence of any indecorum. Does not, then, the advancing and purified civilization of the day demand that colleges should prove their ability to rise to the level of deserved emancipation from sumptuary laws, rather than that, by a retrograde policy, professional schools of any kind should be lowered to the level of involuntary subjection to such laws?

But we contemplated a closing reflection to this section, as follows:

It may be questioned whether, with our familiarity with the advantages of the present, and our comparative incapacity to realize, as by experience, the disadvantages of the past, we duly appreciate the bearings of the great contrast between them. Consider, then, that classical instruction, not essentially different from the present, dates back to days when those mighty agencies of popular enlightenment and kindly civilization — the public school; the popular lecture; the cheap, ever present, and well-filled, periodical; the free library; the wide extended and diffused facilities for cheap and rapid travelling, so influential in opening and liberalizing the mind; the Sunday school, too, and generally accessible kindly and helpful pulpit ministrations, sources of intelligence as well as of moral and religious soundness — when all these, were nearly or quite unknown. In a word, the truly educating agencies of civilized practical life, were far more meagre in earlier days than now. Hence many a bright and steady lad, of twelve to fourteen years, now,

could far exceed in mental development, and general ability to
act in current life, many a rude bumpkin of former days.
Hence also — and this is a point not often considered, as would
appear — so large a proportion of one's total education being
accomplished by the common and constant agencies of ripening
civilized society, a less proportion is left to be still committed
to special organizations expressly designed to impart instruction. Therefore, there seems to be no need for the general or
technical scientific school to be sensitive about adopting as the
total time appropriated by them, the stereotyped allowance of
six or seven years, as in the usual classical course of four years,
followed by a professional course of two or three years. Indeed
a general and technical course, united, of from four to six years,
added to what the best public schools and academies can now
do for diligent members of them, would doubtless place their
recipients more than on a par in general culture and available
power, with the graduate, in generations gone by, of such a
seven years' course as could then have been had. If, then, a
seven years' course be still retained, as the ideal of a full extent
of general and professional school training, it would be with a
view to greatly raising the standard of both general and professional scholarship, over that of times when the school was far
less richly supplemented by the educating agencies of common
life than now. Such a result is most desirable, in behalf of still
continued human progress, while the enlarged area of knowledge offers ample resources for filling seven years of time with
elevated, delightful, and fruitful study. Meantime, we see in
these efficiently educating instrumentalities of our enriched
modern life, so many of which are especially consonant with
scientific study, a source of *that substantial equivalent for the
old collegiate disciplinary preparation for professional study*, which
the technical schools have, at present, partly to rely upon.

7

V.

Conclusion.

Sources of information concerning polytechnic instruction in Europe are remarkably, and unfortunately, scarce and inaccessible. Long extended encyclopœdia articles on education, superior institutions of learning, and nations, in Europe, pass over the polytechnic institutions, which there justly claim equality of rank with the highest, with bare allusions, or partial enumeration; quite barren of all definite information. This may arise from the comparatively recent origin of these schools, whereby they have not yet fallen into a recognized place in national systems of education. In view of the probable lack of information still remaining in various quarters, concerning the number and character of European polytechnic schools, we have thought that the best concluding section of these notes would be a brief account of some of them, and notes of matters suggested by a view of them, as follows:

In France. The *Imperial Polytechnic School.* This celebrated institution was founded in 1794. Its course of study occupies but two years, but this is only because its requirements for admission, especially in mathematics, would be a fair qualification for a professorship in many institutions, while its own professors have often been the generally acknowledged leaders in their respective branches. This school being, moreover, mainly one of general science, it is supplemented for purposes of strictly professional and technical education, by various special schools, some of which are the following:

The School of Roads and Bridges, for the special training of civil engineers. Course three years.

The National School of Mines, with ample illustrative collections, and a course covering three years, for the professional training of mining engineers.

Three National Schools of Arts and Trades, in conjunction with the splendidly equipped *Conservatory of Arts and Trades at Paris*, form an effective instrument for educating higher artizans.

The Imperial School of Forestry.

The Imperial School of Agriculture.

All these high and useful institutions, and others like them, are as yet, being of so recent origin, out of the pale of the great central state department of National education, known as the " University of France," and which embraces the whole old and long organized graded system of National instruction, from the primary schools to the Academies, so called, which are under the charge of eminent Faculties, and have a university character.

The above institutions are, however, national ones, but there is one, the *Central School of Arts and Manufactures*, which is a private institution, founded in 1829, of too high grade to be overlooked. Its courses occupy three years, and provide for the wants of Civil Engineers, Mining Engineers, Mechanical Engineers, and Chemical Technologists.

IN GERMANY. Here, as might be supposed, from the reflective turn of the German mind, national education is more thoroughly organized than anywhere else in the world, and popular education, through common schools, more universal than even in this country, except perhaps in the most favored portions of New England.

The comprehensive organization of German schools, of all grades, is as follows :

Primary.

All the Elementary Schools.

Secondary.

Classical Schools ; Real Schools ; Artizan Schools.

Superior.

Universities ; Polytechnic Institutes.

The Classical schools, called gymnasia, are of about the same grade as our classical colleges. The Real schools are about equivalent to the parallel "scientific courses" advertised in some of our colleges, where physical and mathematical studies, with modern languages, largely replace attention to sundry frivolities of pagan mythology. The Artizan schools, or industrial colleges, are yet more decidedly modern and practical, and stand in a relation to the Polytechnic Institutes, or Industrial Universities," similar to that of the Classical Schools (colleges) to the old Universities. In 1852, there were 26 of these industrial colleges in Prussia, and their substantial equivalency to the classical schools, and our own colleges, is seen in the fact that there, as here, fourteen years is the minimum age for admission to them, while the actual age on entering is considerably higher.

Coming now to the true Polytechnic, or Professional Institutes, we find, among others :

The *Royal Trade Institute* of Berlin, founded in 1821, with a general course of three years, followed by three special courses, for civil and mechanical engineers ; for professional chemists ; and for architects.

The Polytechnic Institute at Vienna was founded in 1815. It includes its own preparatory (real school) course, of two years, followed by a technical course of five years, also a commercial one, and commanding a total attendance upon its regular courses, of 1637 students in 1852.

The Bohemian Nobles' Technical Institute at Prague, founded in 1806, with a preparatory course of two years, and a technical course of three years.

In *Bavaria*, also, there are twenty-six of the artizan or trade schools (industrial colleges) having courses of three years each, preparatory to the three superior polytechnic schools, the oldest of which is the *Polytechnic School at Munich*, founded in 1827. It embraces a preparatory course of three years, and a polytechnic course, proper, of four years.

The technical schools of *Saxony* are of a high order, embracing in their lower grades, the *Royal Trade and Building School* at Chemnitz, with courses respectively of four, and two, years. Above these, are the *Royal Polytechnic School* at Dresden, with a lower and upper section, embracing courses of three, and two years, respectively. Also the celebrated *Mining Academy* at Freiberg, the oldest in the world of its kind, which was founded in 1765, and provides a four years' course of study.

The Polytechnic School at Carlsruhe in Baden, established in 1825, is remarkable for its completeness of organization, embracing a foundation course of three years, followed by numerous technical courses, viz.: one in Engineering, of three years; in Architecture, of four years; in Technical Chemistry, of two years; in Mechanism and Technology, of two years; in Forestry, of two years; in Commercial Science, of one year; and in Postal service, of two years.

GREAT BRITAIN. While this nation was fancying itself to be secure in its commercial and manufacturing supremacy, the London Exhibition of 1851, roused it to a sense of the danger of its falling into a secondary scientific industrial position, owing to its comparative neglect of Modern Applied Science in its higher schools of learning. Glasgow University, however, in 1839, Kings' College, London, and Queen's College, Birmingham, in 1851, were giving formal and quite elevated theoretical and practical instruction in Applied Science.

King's College embraced courses of three years in civil and mechanical engineering, and in general and technical chemistry, requiring sixteen years as the age for admission.

Queen's College announced courses in civil engineering and architecture of three years duration, requiring their entering members to be eighteen years of age.

There are also, in London, we think, a College of Civil Engineers, a Government School of Mines, and a Department of Science and Art in the Institute of Civil Engineers; besides numerous Schools of Industrial (ornamental) Design throughout the United Kingdom, and a College of Civil Engineers for the Indian department, at Madras, India, notices of which we have met in Madras papers; and, without doubt, means must exist—in scientific chairs of instruction attached here and there, to the other colleges and universities, and supplemented, perhaps, more than elsewhere, by private study, or by the adoption of continental precedents ready furnished to hand, or by attendance at continental schools—for educating the accom-' plished engineers, to whose qualifications, however attained, British engineering works testify. We therefore close this notice of foreign polytechnic institutions, with the remark, that the one at Carlsruhe is the most nearly typical one, from its comprehensiveness of organization.

The preceding statistics may be presumed to be interesting, if only as showing what earnest and intelligent fellow laborers have done and are doing elsewhere, and under different political systems from ours. But they serve a higher end. They demonstrate the existence of a universal demand, in all civilized countries, for a new form of general educational culture, and professional training; not to supplant the old, which includes much that is permanently precious, but to run parallel with it, as the legitimate outgrowth of modern science and life, and as the fountain of supply for the new order of intellectual and industrial wants.

This view is confirmed by the fact that the continental appreciation of polytechnic instruction is such, that the larger and lesser European States make appropriations for its support within their borders, as regularly as our American States do for common school instruction.

Some may have a conceit that the man-developing effect of freedom alone, without special educating organizations, is an equivalent to the elaborate systematic instruction, thought of, perhaps, as only necessary to counterbalance the repressing agencies of despotic governments. But with duly admiring deference to Yankee ability to fall back upon native resources in many an emergency, we think the following to be, rather, the true line of argument, relative to this point. If the numerous and crowded polytechnic schools of Europe accomplish so much, as they indisputably do, with all the depressing hindrances of a half-suffocated civil life as the political lot of their graduates, what might they not do, if every graduate was there, as in this country every person is, one of the royal family? In other words, if partly untutored American freedom can compete with the world besides, in many of the truly best contributions to World's Exhibitions, and well-called "Universal Expositions," what might not thoroughly cultured and trained American freedom accomplish, with its fire and elasticity acting through *finished* intellectual machinery, such as thorough scientific and polytechnic education may produce out of the material, turned out in an only partially wrought form by the common school from the native ore of original talent?

Finally, therefore, it is to be most earnestly hoped that at least among the institutions, having so large resources as those provided for by the National land grants to the States for endowing Scientific Institutions in each, especially if also otherwise liberally endowed, if not among the riper Technical Schools of this country, some one will ere long be found, to signalize an era in American scientific education, and confer a new and peculiar glory on the fortunate State containing it, by constituting itself a true typical *Polytechnic University*, characterized by a completely comprehensive unity of design, and built up, if gradually, not in a disjointed manner, but, even in the planning of its grounds and distribution of its buildings, as well as in its component courses, and "schools," in accordance with a complete original plan.

Such a "University" should be distinguished—First: by a central foundation, or general, scientific school, of high character, with a course of liberal training in general disciplinary and useful knowledge, embracing such a proportion of elective studies as to possess due flexibility in providing for the wants of those who should be contemplating any particular subsequent technical and professional course. Second: it should be distinguished by possession of the highest true university attribute, of making express provision for the indefinitely extended pursuit of single or associated subjects of general science, and *real* learning. Third: circling as it were, around this central general school, which should be in a plain, but rich and massive structure, there should be a collection of all the technical professional schools, congruous with the distinctive idea of a *Polytechnic*, rather than a Humanistic, University, viz: one of *Civil and Topographical Engineering* (sections of one school); one of *Mechanical Engineering;* one of *Mining and Metallurgy;* one of *Civil Architecture*, naval included; one of *Technical Chemistry;* one of *Physical Technology and Technical Natural History* (sections of the proper school of "arts and trades"); one of *Agriculture and Forestry;* one of *Industrial Ornamental Design* (Schools of Purely "Fine Art" should, we think, collectively form a separate "*Art* University" disconnected from the distinctively "*Industrial*" or Polytechnic one); a *Commercial* one of high order; and a *Technical Normal School*, for the training of professors of general or technical science. Fourth: As a collateral group of buildings, each to be as far as possible an architectural model, there should be the General Museum and Assembly Hall, the General Library, the Chapel and Observatory. Fifth: The plan should include Professors' residences and Students' homes, the latter to accommodate six to twelve persons each, with the householder's family; a gymnasium, and the requisite lodges. Also, in respect to grounds, they should be ample enough to embrace, wood, lawn, ground for manly field games; a botannical garden, and arboretum; and a park and pond for animals.

Lastly, the buildings of the technical schools, should include the various laboratories, cabinets, scientific society rooms, apparatus and work rooms, appropriate to their uses.

It would be easy to add the outline of a simple plan of distribution for all the foregoing structures, by which the essential unity of the entire establishment should be elegantly, as well as visibly, expressed in the very arrangement of its material components. But we forbear, and pass on to consider briefly the subject of the *Endowment of Polytechnic Schools*. Colleges are quite generally, and not incorrectly, regarded as existing for the general intellectual, and, incidentally at least, for the moral good of the entire country. They exist for this end more than for any merely private, especially any pecuniary, good of their members. Hence they are treated as having a recognized claim upon the wealthy liberality of the country, and are very often quite largely and cheerfully endowed, as may be seen by the frequent large donations to them, reported in the newspapers at " commencement " times.

Professional Schools, however, especially those of Law and Medicine, while existing in a very high sense for the general good, exist, to a greater comparative extent than colleges, for the immediate pecuniary benefit of their members. They are, therefore, except Theological schools, less generally and liberally endowed, and more supported by current tuition receipts. But the exception shows that a school should not go unendowed merely because a professional one. Let us, then, examine the claims of Polytechnic Schools in reference to this question of endowment.

We should confess the impropriety of publishing, here, definite statistics as to the endowments of the schools given in the Table in Section I, but it may be said, in general terms, that they vary, from sums too small to name, up to $50,000, $100,000, $250,000, $750,000, $1,000,000, and upwards. And the life of the institutions, thus variously conditioned, may be supposed to vary correspondingly, from that of a dry and wiry cedar growing in a cleft of a rock, drawing support from everywhere but the immediate place of its growth, to the spreading luxuriance of willows by the water courses. But, seriously, the Polytechnic Schools provide a ready entrance to lucrative positions for their graduates. Still, the labors of those graduates *tend directly and powerfully to increase the wealth of the nation*, by developing its mineral resources; by opening up avenues of inter-communication, as in railroads, canals, and river and harbor improvements; by adding to its mechanical appliances; and by the increased production of articles of commerce derived by application of Industrial Physics, Chemistry, and Natural History to many 'arts and trades. On the other hand, the studies of Polytechnic Schools, being largely material, *require elaborate material appliances for their most successful prosecution;* Models, Instruments, Apparatus, Cabinets, Botani-

8

cal Gardens, and Scientific Libraries, with numerous Diagrams, Illustrative Drawings, and Charts. They thus have a two-fold claim to a liberal endowment, at least with funds to equip them handsomely with these necessary material appliances, if not with endowed professorial chairs.

But there is another fund which Polytechnic Schools especially need, viz: a publication fund. Being partly, at least, a unique class of schools, their text-books can often best be prepared by their own professors. The cost of making such books is necessarily great, and their sale of necessity relatively small. Hence, as it is by no means an unknown custom, such works should be published, in part certainly, from a fund for the purpose.

We here, though rather abruptly, close, considering that, if these Notes have not failed of their immediate object, they have justified their title page, in that they have shown that Polytechnic Schools are, in their *nature*, truly professional ; that their *position* is, provisionally, and in part, one of compromise with their ideal condition ; that their *aim* is, to attain the everywhere undisputed rank of fully professional schools ; and that their *wants* are, adequate preparatory schools, (colleges) which, in turn should have previous academy training courses of general science; and material detachment from collegiate and professional schools of the humanistic type—not, of courses in any narrow exclusiveness of spirit, but as a matter of expediency. Our work thus done, we only add a word of ancient testimony to the impossibility of knowing the whole of anything, much less, of everything, and hence, to the propriety of the recognized double line of learned pursuit, humanistic, and polytechnic, which we have advocated. In this testimony, the great regal example of the polytechnic learning and practice of old, who says, " I gave my heart to search out by wisdom concerning all things that are done under heaven," and, " I made me great works," declares :

" *He hath made everything beautiful in his time : also He hath set the world in their heart, so that no man can find out the work that God maketh from the beginning to the end.*"

OTHER WORKS

BY THE SAME AUTHOR, ON

Practical and Descriptive Geometry and Geometrical Drawing.

Published by JOHN WILEY & SON, 535 Broadway, N. Y.

ELEMENTARY COURSE.

1.—

2.—

3.—ELEMENTARY PLANE PROBLEMS. (*In preparation.*)

4.—DRAFTING INSTRUMENTS AND OPERATIONS. Div. I.—*Instruments and Materials.* Div. II.—*Instrumental Operations.* Div. III.—*Constructions in two dimensions.* Div. IV.—*Elementary Æsthetics of Geometrical Drawing.* Price $1.25.

5.—MANUAL OF PROJECTIONS, ETC. Div. I.—*Elementary Projections.* Div. II.—*Details in Masonry, Wood and Metal.* Div. III.—*Elementary Shades and Shadows.* Div. IV.—*Isometrical Drawing.* Div. V.—*Simple Structure Drawing.* Price $1.50.

6.—ELEMENTARY LINEAR PERSPECTIVE. Part I.—*Primitive Methods.* Part II.—*Derivative Methods.* Price $1.00.

HIGHER COURSE.

7.—

8.—DESCRIPTIVE GEOMETRY. *General Problems.* Price $3.50.

9.—

10.—

The above published works are all fully illustrated with cuts and plates. They are used in several of the Scientific or Polytechnic Schools of the Country; and have received warm commendation in various quarters. The volumes of the Elementary Course, are especially adapted for the upper classes in High Schools and Academies; and for the Scientific Undergraduate Courses in Colleges, as well as for the lower classes in the Polytechnic Schools; and for the self-instruction of Artizans, etc. Vol. 6 is also especially adapted for Ladies' Seminaries and Schools of Design, in which the principles of perspective are taught.

www.ingramcontent.com/pod-product-compliance
Lightning Source LLC
Chambersburg PA
CBHW031747090426
42739CB00008B/914